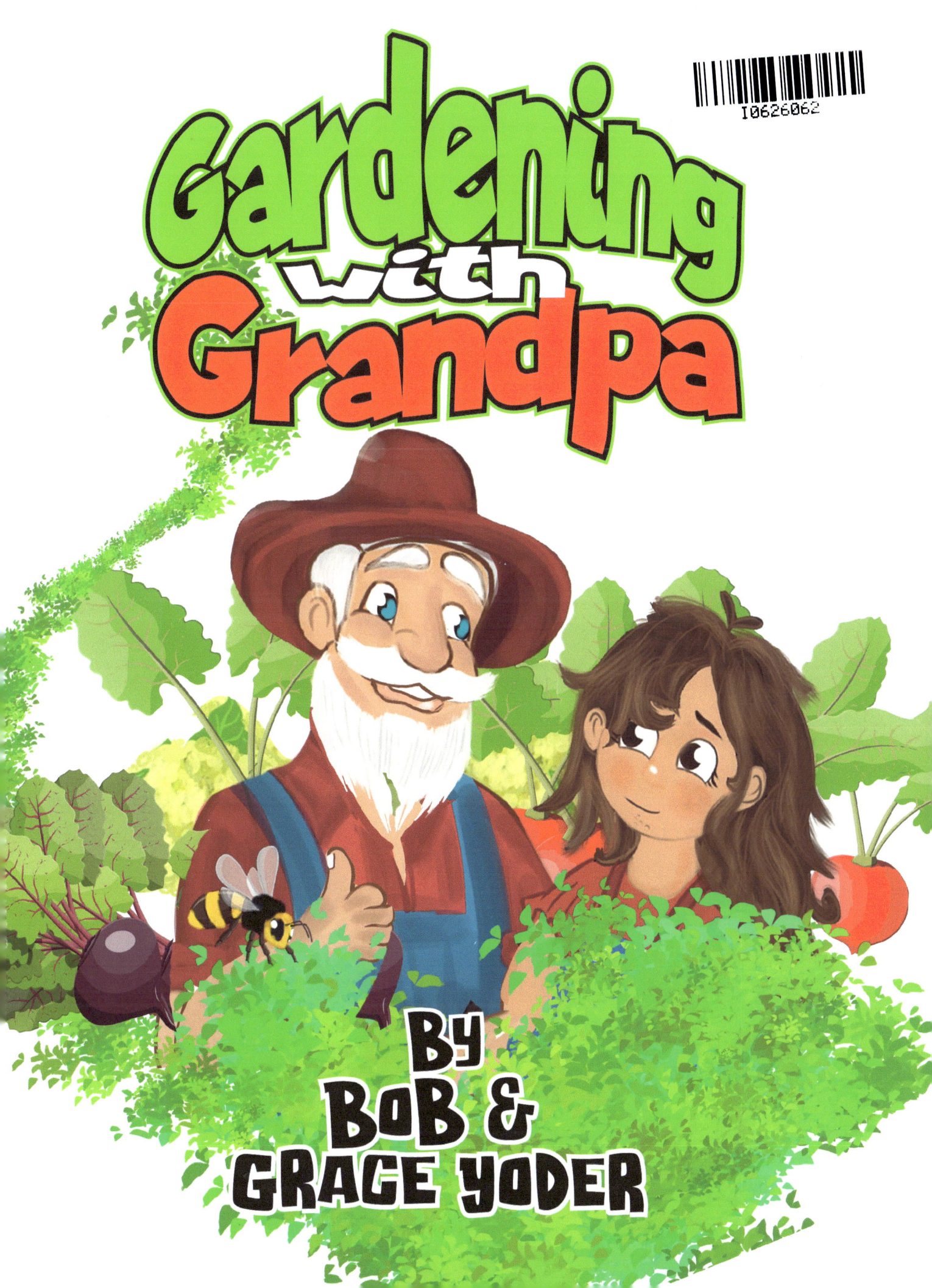

Gardening with Grandpa

By BOB & GRACE YODER

DEDICATION

To Grace Yoder, my
granddaughter, thank you for sharing your
ideas and helping me write this book.
Your thoughts are included in this book and
will, hopefully, inspire, educate, and help other
kids plant gardens with their grandpas.

Selecting the right location
for our garden.

Preparing the soil with a
hoe or rototiller.
Always use recyclable food
for a great fertilizer.

Discuss with each other what we would each like to plant and go to the local market to get seeds to plant.

Make the rows with a **Hoe** or by **Hand**.
Plant and water the seeds.

Continue to take
care of your
garden and enjoy
the beautiful farmland.

Monitor the weather for **Sun** and **Rain** to help your plants grow. Always watch for weeds.

Helping **Grandpa** weed and keep the garden organic.

Plants are now growing, and it is so fun watching the tiny flowers grow into big fruits and vegetables.

The bees are our friends and help pollinate the plants.

It's getting time to pick and harvest all of our plants.

We have to rototill our garden so that it can rest over the winter and be ready for next spring.

We can now can or freeze our vegetables for the winter months.

We always take some of our harvests to the local food bank or blessing boxes.

We enjoy going in the kitchen and helping each other make our favorite recipes.

It's been a long, hot summer and a lot of work but it sure is worth it when our cupboards and freezers are full of what we planted.

We are now getting ready for next year's garden with more fresh ideas.